Blast Off with Dadiso

★ Science, Secrets, and Surprises on the Moon ★

A STEM Story about NASA's Artemis Missions

Blast Off with Dadiso

Science, Secrets, and Surprises on the Moon

A STEM Story about NASA's Artemis Missions

by Rudo Kashiri

Curious Minds Explorers Publishing

A Curious Minds Explorers picture book

Ages 8-12

Blast Off with Dadiso

Science, Secrets, and Surprises on the Moon

A STEM Story about NASA's Artemis Missions

Illustration by Roshanay

Published by Curious Minds Explorers Publishing

Stories That Launch Curiosity

ISBN: 979-8-9996203-7-8

Printed in the United States of America

This book is a work of fiction inspired by real science. Names, characters, places, and events are used to encourage curiosity and learning and are presented in a fictionalized format.

This book is your **Mission Log** for an epic adventure to the Moon.

As you read, you will:

- Learn why NASA is heading back to the Moon
- Discover what Artemis is and how astronauts will travel
- Explore the Moon's South Pole and the mystery of water ice
- Imagine what life could be like inside a Moon habitat
- Take on a STEM Splashdown Parachute challenge

When you see a **Mission Log** question at the bottom of a spread, pause and answer it.

Grab a pencil. Fuel up your imagination. Your mission starts now.

It's time for liftoff!

This book is dedicated to my granddaughter, Dadiso. May you always stay curious, ask brave questions, and dream boldly whenever you look up at the stars.

The sky is not your limit; it is only your beginning.

"The future belongs to those who believe in the beauty of their dreams."
— Eleanor Roosevelt

Welcome, Explorer!

When you look up at the Moon, what do you imagine?

Some people see a bright light in the night sky.
But for curious kids like Dadiso and maybe you, the Moon is a place full of mysteries, questions, and dreams.

This story isn't just about going back to the Moon.
It's about how **you** could help shape what comes next.

Are you ready to join Dadiso and her grandmother, Ambuya, on a journey to discover what comes next?

Grab your Mission Log. Fuel up your imagination.
Liftoff begins now.

REACH
FOR THE
STARS
THE
MOON
OUR GATEWAY
TO DISCOVERY

Dadiso sat cross-legged on the rug as the Artemis rocket's engines rumbled through the TV speakers. On the screen, the countdown numbers glowed bright enough to light up her face.

In the dark reflection of the screen, she pictured herself beside the rocket, heart thumping, ready for adventure.

On the couch nearby, Ambuya sat in a patch of warm sunlight, knitting quietly, click, click like a steady countdown of her own.

"Ambuya," Dadiso asked, eyes wide, "didn't we go to the Moon before?"

Ambuya smiled and set her knitting aside. "Yes, we did. But this time, we're going back to learn more, and to stay longer."

Dadiso hugged her knees. "So people might really live on the Moon?"

"We hope so," Ambuya said. "We need to learn how to live on another world, and the Moon is where we can practice. Astronauts and scientists will explore and discover, and they'll learn to use what the Moon offers, like water ice. That could help people stay longer and travel farther into space."

Dadiso pictured the dusty ground beneath her boots, one small hop, then another. "Do you think I could go someday?"

Ambuya squeezed her shoulder. "With your curiosity and determination, anything is possible."

MISSION LOG ☆

If you could visit the Moon, what would you want to explore first? ♡

Ambuya reached for a worn scrapbook from the shelf. When she opened it, the pages smelled like paper, time, and memories.

Dadiso leaned closer. Inside were black-and-white photos of astronauts, towering rockets, and a flag planted in dusty Moon soil.

"This is Apollo," Ambuya said softly. "It's what we call the missions when people first walked on the Moon."

Dadiso traced a picture with her finger. "They really stood there...on the Moon."

"Yes," Ambuya said. "Apollo helped us learn what the Moon is like. But Artemis will help us learn how to return and stay longer."

She tapped the page gently.

"Apollo visited. Artemis will return to build and explore."

Dadiso's eyes sparkled. "So Artemis is like the next chapter?"

Ambuya nodded. "A new chapter for a new generation."

Dadiso smiled. "Maybe one day I'll have my own mission patch to add to a scrapbook."

Ambuya's smile widened. "Maybe you will."

OUR GATEWAY
TO DISCOVERY
My Space Ideas
DID YOU KNOW?
Neil Armstrong was the first person to walk on the Moon (1969).
Gene Cernan was the last (1972).

Dadiso glanced back at the TV. "Why is it called *Artemis*?"

Ambuya chuckled. "In old stories, Artemis was Apollo's twin sister. Apollo was the name of the first Moon missions, so NASA chose Artemis for the missions that are returning to the Moon."

Dadiso imagined the two names Apollo and Artemis shining like story-stars, pointing the way back to the Moon.

"So Artemis is NASA's plan to go back to the Moon?" Dadiso asked.

"Yes," Ambuya said. "Artemis is a program that will send astronauts around the Moon and back down to the surface. It will also help us learn what we need to explore deeper into space."

Dadiso sat up straighter. "That sounds huge!"

"It is," Ambuya agreed. "And big missions happen step by step."

Ambuya's voice grew proud. "Artemis also aims to land the first woman and the first person of color on the Moon."

Dadiso smiled, imagining a fresh footprint in the Moon dust. "Maybe someone from my class could go."

Ambuya nodded. "Maybe. And maybe someday... it could be you."

DID YOU KNOW? ☆

Each of the six Apollo lunar landing missions (Apollo 11, 12, 14, 15, 16, and 17) sent two astronauts to walk on the lunar surface.

Dadiso scooted closer, practically bouncing. "Okay, Ambuya, when does the Moon adventure actually begin?"

Ambuya opened a calendar covered in stars and tiny rockets. She held up three fingers like a countdown.

"Think of Artemis in three big steps."

"Step one," she said, folding down one finger.

"Artemis I tested the giant rocket and spacecraft with no astronauts. It was a super-important practice run to make sure everything worked."

"Step two," Ambuya continued, folding down a second finger.

"Artemis II carried astronauts around the Moon and back home again."

Dadiso's jaw dropped. "Around the Moon?! Like... whoooosh all the way around?!"

Ambuya grinned. "Yes, around the Moon. NASA practiced with a real crew to make sure Orion and everything inside worked safely, then they came home to tell the story."

Ambuya folded down her final finger.

"And step three: Artemis III, the big moment, boots on Moon dust!"

Dadiso's eyes shone. "So it's a plan: test, fly, then land!"

"Exactly," Ambuya said. "Big missions happen one careful step at a time."

★ THE JOURNEY AHEAD ★

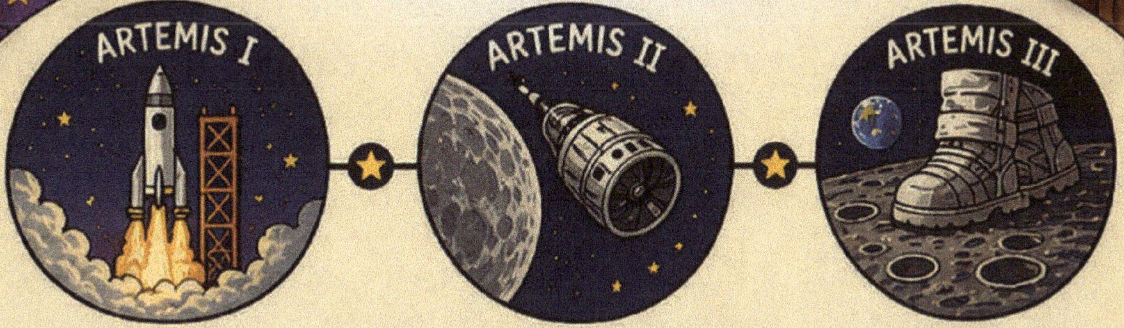

2022	2026	2027+
ROCKET	ASTRONAUTS ORBIT THE MOON	ASTRONAUTS LAND ON THE MOON

★ ARTEMIS AT A GLANCE ★

Artemis I: Test flight

Artemis II: Astronauts orbit the Moon

Artemis III: Astronauts land on the Moon

Ambuya tapped her tablet. "Want to see that launch again up close?"

"Whoosh!" A glowing rocket popped onto the screen.

Dadiso pointed so fast her finger nearly smudged the glass. "That rocket is ENORMOUS! It looks like it could poke the Moon!"

"That's the Space Launch System, or SLS," Ambuya said. "It's one of the most powerful rockets NASA has ever built, strong enough to lift a spacecraft off Earth and send it racing toward the Moon."

Dadiso leaned in. "Okay, but where do the astronauts actually sit? Is there, like, a driver's seat?"

Ambuya zoomed in near the top. "In Orion," she said. "It's the crew capsule, like a tiny space home that carries astronauts through space with everything they need for the trip."

Dadiso squinted at a pointy piece above Orion. "What's that little tower for?"

"That's the Launch Abort System," Ambuya said. "Think of it like a rocket-powered safety helper. If there's trouble during launch, it can pull Orion away fast so the astronauts can get to safety."

Dadiso nodded, building the rocket in her mind like a stack of space blocks. "So SLS is the muscle... Orion is the crew's space home... and the abort system is the safety saver."

Ambuya beamed. "You've got it: three parts, one Moon-bound team!"

Dadiso tilted her head. "But the Moon is *huge*. Where do you even park a spaceship?"

THE SLS ROCKET
A POWERFUL ROCKET BUILT FOR DEEP SPACE EXPLORATION!
LAUNCH ABORT SYSTEM
Protects astronauts by quickly pulling Orion away from the rocket in an emergency.
ORION SPACECRAFT
The astronauts' space home. It takes them to the Moon and back safely.
CORE STAGE
Holds the fuel that powers the rocket and its engines.
SOLID ROCKET BOOSTERS
Provide extra power at launch. They burn fast and help lift SLS off the pad!
NASA
DID YOU KNOW?
SLS is taller than the Statue of Liberty.
Orion is the astronauts' space home.

"At the South Pole," Ambuya said, turning her tablet so Dadiso could see a silvery Moon map. She traced a circle near the bottom with her fingertip.

"Right about here. Imagine putting a sticker on the Moon!"

Dadiso blinked. "The Moon has a South Pole?"

Ambuya laughed. "It does! And near the South Pole, there are craters that never see sunlight." She lowered her voice dramatically. "They're like giant bowls of shadow, Moon freezer caves. The bottoms stay freezing cold all the time."

Dadiso hugged her arms. "Brrr! Why go somewhere *that* cold?"

"Because those shadowy craters may be hiding something amazing... water ice," Ambuya said. "Treasure, but the kind you can drink!"

She tapped the screen. "Astronauts could melt it for water, and it could even help make oxygen to breathe and fuel for rockets."

Dadiso gasped so hard she squeaked. "Water... on the Moon?!"

"Yes," Ambuya said. "Past missions found clues that it's there. If we can get to that ice, it could help future explorers stay longer."

Dadiso's imagination raced. "So water is valuable on the Moon, too?! Wow!"

"Exactly," Ambuya said. "That's called using local resources, finding what you need right where you are." She winked. "Like making a snack with what's already in your kitchen... but on the Moon."

DID YOU KNOW?

Some parts of the Moon's South Pole are always in shadow and never see sunlight!

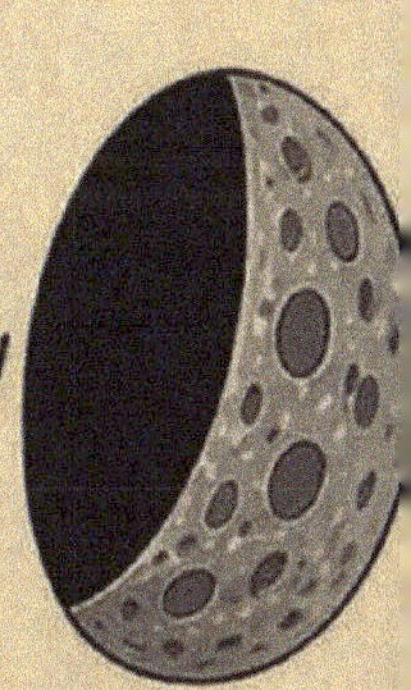

Dadiso leaned closer as Ambuya scrolled on the tablet. "So... what do astronauts *actually* do once they get there?"

"They do science," Ambuya said. "They collect Moon rocks, study the dusty ground, and look for clues about how the Moon formed, almost like reading a very old story written in stone."

Dadiso pictured an astronaut kneeling in the dust with a toolkit and a sample bag. "Moon detectives!"

Ambuya smiled. "Exactly. They'll set up instruments, test new tools, and run experiments."

Dadiso leaned closer. "Will they build things like a Moon lab... or a Moon garage?"

"Yes," Ambuya said. "They'll test ways to build and repair equipment, and to explore safely so future missions can stay longer and do even more."

Dadiso's voice dropped to a whisper, like she was sharing a secret with the stars. "I want to discover something no one has ever seen."

Ambuya nodded. "That's the heart of science."

Dadiso glanced up. "And all of this... is it to get ready for Mars?"

"Yes," Ambuya said. "The Moon is like our practice field."

Dadiso's grin returned. "Then we might find something brand-new!"

Ambuya nodded. "That's the hope."

DID YOU KNOW?

On the Moon, you weigh about one-sixth as much as on Earth.

MISSION LOG

Which astronaut job would you want to try, and why?

____

____

____

____

Dadiso scrolled on the tablet and stopped at a picture of a Moon habitat, little rooms tucked together like space blocks. "Where will they sleep?"

"In special habitats," Ambuya said. "Tiny Moon homes protect astronauts from scorching heat, freezing cold, space radiation, and sharp, clingy Moon dust."

Dadiso studied the picture. "So... not like camping with a sleeping bag?"

Ambuya chuckled. "Moon living takes careful design: air to breathe, water to drink, power for lights and tools, and a safe place to rest when your body says, 'Recharge!'"

Dadiso's eyebrows lifted. "And what about walking outside?"

"They'll wear spacesuits," Ambuya said. "And they'll move differently, too. Moon gravity is much lower than Earth's, so each step can turn into a hop, like slow-motion jumping!"

Dadiso grinned. "I would hop everywhere. Boing! Boing! Boing!"

"It does look fun," Ambuya agreed. "The Moon is a tough place to live."

Dadiso paused. "Okay... but what about food?"

"Astronauts may grow plants inside the habitat," Ambuya explained.

"A garden on the Moon?" Dadiso gasped. "What would they grow first?"

Ambuya smiled. "Maybe tomatoes, lettuce... and strawberries someday, foods that can grow in small spaces and help keep explorers healthy."

Dadiso giggled. "Moon strawberries! Okay, I'm officially moving in."

★ MISSION LOG ★

What would you grow in a Moon garden?

____

____

____

Ambuya tapped her tablet again. "Ready to see the part where Orion comes home?"

A video filled the screen: Orion streaking toward Earth, wrapped in a bright, flickery glow like a comet diving through the sky.

Dadiso's mouth fell open. "Is it... on fire?"

Ambuya smiled. "It looks like fire, but it's mostly super-heated air glowing around the capsule. Orion is moving so fast during reentry that the air around it heats up and glows."

She pointed to the flight path on the screen. "Reentry also has to be just right. If Orion comes in too steep, it can get dangerously hot, making the ride much rougher. If it comes in too shallow, it can skip off the upper atmosphere like a stone skipping on water and not slow down enough."

Dadiso leaned closer. "So how does it not melt?"

Ambuya zoomed in on the bottom of the capsule. "See this? That's Orion's heat shield, a tough protective layer designed to take the worst of the heat, so the astronauts stay safe inside."

Dadiso thought of helping in the kitchen, gripping a hot pan with an oven mitt. "So the heat shield is like a giant space oven mitt!"

Ambuya laughed. "Exactly. Reentry is one of the hardest parts of the whole mission."

Dadiso's voice went quiet, "Going to the Moon is incredible... but coming home safely is incredible too."

Ambuya swiped to the next clip. On the screen, Orion drifted beneath giant red-and-white parachutes, floating down, slow and steady, like a dandelion seed riding a breeze.

Dadiso grinned. "Parachutes!"

"Yes," Ambuya said. "After reentry, the parachutes open to slow Orion down even more and guide it gently to the ocean, like a careful landing on water."

Ambuya added, "Orion is built to float, so it can bob on the waves until the recovery teams reach it."

The capsule hit the waves with a thump and a towering spray of saltwater.

"SPLASHDOWN!" Dadiso shouted, laughing.

Ambuya nodded. "And then comes recovery. Teams wait until Orion is safely in the water, then they head out in small boats to check the capsule, attach floatation gear, and help the astronauts out."

She added, "Helicopters can also help during recovery. They can bring teams and equipment and sometimes transport astronauts once they're safely out of the capsule."

Dadiso watched boats circle Orion like ducklings around their mom. "So even the ocean is part of the mission."

"It is," Ambuya agreed. "A space mission doesn't end the moment you return to Earth. It ends when the crew is checked, cared for, and truly safe at home."

Dadiso hugged her journal. "I like that. Explore bravely... and return safely."

MISSION LOG ☆

How do parachutes help Orion land safely?

♡

Dadiso and Ambuya stepped outside and looked up at the Moon. Dadiso turned to Ambuya. "So... after the Moon, where do we go next?"

Ambuya's eyes twinkled. "That is a wonderful question. The Moon is just the beginning. Some scientists and astronauts are already dreaming about sending humans to Mars."

Dadiso's eyes widened. "Mars?! That's *so* far!"

"It is," Ambuya agreed. "That's exactly why the Moon matters. It's the perfect place to practice staying safe, using local resources, and working as a team on long missions, before we try an even bigger journey."

Dadiso imagined a path: Earth... Moon... Mars. It was like stepping-stones across space. "So the Moon is training for the bigger adventure," she said.

Ambuya nodded. "Yes. Every mission teaches us something we'll need for the next one."

Dadiso held her Mission Log tight. "Then I'm going to keep learning. I want to be ready when it's my turn."

Ambuya squeezed her hand. "With your curiosity and big dreams, anything is possible. The future needs explorers like you."

Night had fallen, and the world outside was still. Dadiso snuggled into bed with her Mission Log open and her pencil poised. Her nightlight painted soft shadows on the walls, turning her room into a tiny space station.

She whispered. "One day, I will explore."

Then she wrote, slowly and carefully:

I will ask questions.

I will learn.

I will help.

Dadiso paused, staring at the page as if it might answer back.

"What will I discover?" she whispered.

From the hallway, Ambuya's voice drifted in, gentle as a lullaby.

"Every big journey starts with a big dream. Keep imagining."

Dadiso smiled and drew one more star in the corner of her Mission Log, small but bright.

"Maybe one day," she thought, "I'll be the one exploring the Moon."

She turned toward the window, where the Moon glowed like a quiet lantern.

And as her eyes closed, Dadiso pictured a future where her curiosity could carry her farther than she'd ever imagined, step by careful step.

☆ MISSION LOG ☆

What kind of explorer do YOU want to be when you grow up?

____

____

____

____

AMBUYA'S CORNER:
Earth, Mars, and the Moon

These three worlds may look like neighbors, but their sizes are very different.

Earth is the biggest and our home. Mars is about half Earth's width. The Moon is about a quarter of Earth's width.

The Moon may be small, but it is a big deal for explorers. It is close enough to reach and a perfect place to practice for bigger journeys.

In science, we learn step by step.

MISSION LOG: If you could choose, which would you explore first, Earth's Moon or Mars, and why?

Mission Patches

Astronauts wear mission patches to show a mission's goals and the team behind it.

Dadiso's patch is a reminder that explorers do more than travel. They learn, build, and solve problems together.

And every journey starts with one powerful idea: **What if?**

Her patch shows a rocket traveling from Earth to the Moon, with astronauts walking on the Moon's surface below. It represents Dadiso's dreams, her questions, and the courage to imagine new possibilities. Because every explorer begins the same way: **with curiosity.**

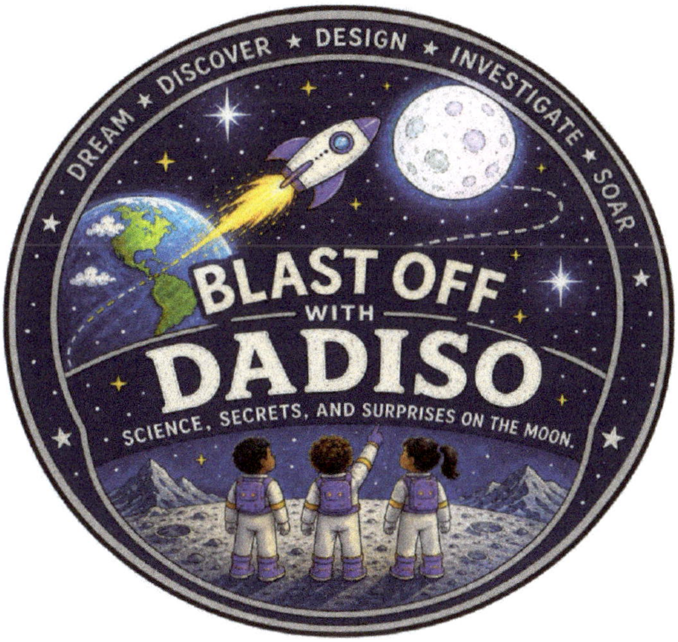

MISSION LOG

Draw your own mission patch.

Choose 2-3 symbols that show what you want to explore.

Author's Note

Hi, I'm Rudo Kashiri, a former NASA educator and the founder of Curious Minds Explorers Publishing.

I wrote this book to help kids see themselves as explorers because space science isn't only for astronauts. It's for anyone who stays curious, asks brave questions, and keeps learning.

The Artemis missions remind us that big dreams happen step by step. And no matter where you live, your curiosity can take you far.

Thank you for joining Dadiso on this Moon mission. Keep wondering, keep exploring, and keep shining.

Stay curious,
Rudo Kashiri

"If you can dream it, you can do it." (African proverb)

SPLASHDOWN PARACHUTE CHALLENGE

Design a parachute that helps your "Orion capsule" land gently in the ocean.

Materials (check as you collect)

☐ 4 strings (10 -12 in each)

☐ Tape

☐ Towel

☐ Scissors (adult help)

☐ Bowl/tub of water ("ocean")

☐ Canopy: coffee filter / plastic bag / tissue paper

☐ Capsule (circle one): ping-pong ball/plastic egg / small paper cup

Step 1: Choose + make your canopy (circle one): coffee filter / plastic bag / tissue paper. Coffee filter: use as-is. Bag/tissue: trace and cut a hand-size circle.

Prediction (circle one): coffee filter / plastic bag / tissue paper

Step 2: Attach strings + capsule. Tape 1 string to 4 spots on the canopy. Tug gently to check they're secure. Gather the loose string ends and tape them to your capsule.

Prediction (check one): ☐ Soft landing ☐ Big splash ☐ Tipped over ☐ Fell too fast

Safety: If you drop from a chair, step stool, or anything high, **ask an adult to help**.

Step 3: Test. Drop into the "ocean," then circle: Soft / Splash / Tipped / Fast

Step 4: Improve. Change ONE thing, then test again.

☐ Bigger canopy ☐ Smaller canopy ☐ New material

☐ Longer strings ☐ Shorter string ☐ Add a float ring

MISSION LOG: Reflection

1) What change worked best, and why?

2) What recovery tool would you add, and why: boat, hook, or float?

www.ingramcontent.com/pod-product-compliance
Lightning Source LLC
LaVergne TN
LVHW070859210926
848482LV00016B/978